Apple Man

by
Jim Holmes

*Living on the Land
in Norfolk
&
Selling on
Yarmouth Market*

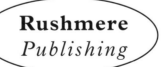

Rushmere
Publishing

First published 1998 by Rushmere Publishing
32 Rushmere Road, Carlton Colville, Lowestoft, Suffolk

Typeset and printed in England by Blackwell John Buckle
Charles Street, Great Yarmouth, Norfolk NR30 3LA

ISBN 1 872992 13 7

Acknowledgements

I am grateful to Percy Trett who has once again allowed me to dip into his wonderful photographic collection and I would also like to thank the families whose names have long-standing connections with Yarmouth Market; the Harveys, Bartrams, Thompsons, Brewers and McCarthys, who have also given permission to use their photographs.

A special word of thanks must go to Dean Parkin at Rushmere Publishing for his help with this book.

Foreword

by KEITH SKIPPER

Jim of all trades - and master of most of 'em! That's how I sum up the little man with boundless energy.

By the time I got to know Jim Holmes he had already become a living legend in the Yarmouth area. This compelling volume makes it clear why he has left distinctive marks in different fields. He calls himself 'an awkward independent' - and that takes into account the occasional flirtation with unpopularity - but I reckon 'hard-working scholar' would do just as well. Jim may have fallen for a few rosy notions of countryside life after the last war, but the way he tackled harsh realities alongside Kathleen, a willing and thrifty wife, underlined a rare ability to learn quickly and to learn well.

He was never scared of long hours, tired limbs and small rewards. Nor did he sulk at the blatant unfairness of it all when cheap foreign imports hit the home market, a truly savage twist to the Ormesby plot. From cacti to chrysanthemums, tomatoes to tulips and bulbs to bees Jim was ever ready to diversify, spreading optimism and enthusiasm like twin composts to feed fresh adventures.

Wonders like Burpee Hybrid, Yunanensis Napsbury, Big Boy, Sprinter, Cupheas Platycentra and the Anzani Iron Horse all became part of a burgeoning career. All the while, Jim was adding to his information bank in several accounts.

An acknowledged expert on apples, he decries the loss of quality and flavour in these, especially Cox's Orange Pippin, and other fruit, vegetables and poultry: "Last time I enjoyed a real tomato was from a plant self-sown growing in the ruins of the old greenhouses that collapsed in a gale."

Jim also relishes the flavour of the colourful characters which his years on Yarmouth Market provided; the sort of cavalcade usually reserved for novelists with fervent imaginations. For starters, you can meet 'Shacky' Bond in a bowler hat and buskins; Freddie Broom in a straw hat trimmed with tomatoes; a scruffy little man who cried 'Fowls are cheap!'; Prince Monolulu, the racing tipster, and the Big Drips who couldn't see as far as the end of their noses.

Humour and shrewd observation are the Holmes' hallmarks as he sets out this memorable stall, and it is no surprise to find how often he was in demand to take part in television and wireless programmes. Jim's wide range of interests, from the Parson Woodforde Society to the restoration of the Yarmouth Fishermen's Hospital, emphasises a constant eagerness to be involved in local affairs and to do something useful with his talents and his time. This book is a telling tribute to the proud achievement of those aims. You make it sound rather elementary, my dear Holmes, but I can't help wondering how on earth you crammed so much in and yet still found time to hear the last corncrake in Ormesby!

Keith Skipper
Cromer, 1998

Contents

CHAPTER 1

Any Fool Can Grow a Lettuce

After six years of war I wanted to continue an active outdoor life, growing things instead of destroying them and, above all, be my own boss. My capital was limited so farming was out of the question but market gardening and fruit growing were just within my modest means and that is what I did. Soon after the war, starry-eyed, inexperienced, eager and impatient as I was, everything seemed easy. Any old produce sold without much effort and I did very well. So well that I got married in 1950!

In 1947 I began market gardening at Ormesby on four and half acres of land planted with mixed fruit and some open ground, the whole being very run down and overgrown with rubbish, especially speargrass. The old cottage was 200 years old and lacked every modern convenience except water and electricity which I had laid on after I bought it. No modern toilet, no bathroom and the kitchen was in an outhouse with an electric cooker, a sink with a draining board and a kitchen cabinet. It had been two labourers' cottages which Bob Leath had made into one but I could not afford more than the basic necessities. It was several years before we could afford to modernise it and when the main sewer arrived in Ormesby we had at last a W.C. and bathroom. We were fortunate in buying an almost new Rayburn stove which our neighbour was dispensing with for just £10 and Dennis Knights installed complete to warm the cottage. It transformed it so we not only had an outstanding cooking stove but a warm cosy kitchen-living room as well.

The orchard and fruit bushes were old, almost a museum, originally a wonderful selection of apples, pears, plums, raspberries,

Our cottage and first greenhouses.

gooseberries, and red, white and blackcurrants, well chosen but long past their prime. They should have been grubbed out years before but had proved a gold mine during the war when fruit was at a premium. We had been told that it was a good living with no difficulty in selling the fruit as greengrocers and fruit merchants called to buy what was available which may have been true during the war but things rapidly changed and by 1947 there was plenty of fruit but no buyers! In desperation we took a stall on Yarmouth market which we retained until we retired in 1978 and sold our produce direct to the public.

For about three years it was a fairly easy time in a false Eden although I was quickly learning the ropes. The crunch came in 1950 when reality faced the industry and the bubble burst. By this time the free movement of horticultural produce had resumed and imported foreign fruit, missed so much during the war years, began to flood the markets and the easy times for British growers were over.

This was brought home to us very sharply that autumn. We had a very large crop of Dr. Harvey apples, hopefully our income until Christmas, and I remember showing my wife the huge crop ripening in our apple store, slowly turning that lovely golden yellow. "There," I said, "that is as good as money in the bank." Poor fool! Little did I think that we would barely sell a stone or two and those mainly to our faithful Dr. Harvey addicts who stayed with us until the trees were finally felled in 1970. The rest we threw out and ploughed in. By this time the public at large tended to ignore local apples, pears and plums after enjoying their fill of foreign produce at high prices just before our fruit appeared.

Well-established holdings did not immediately feel the effect, being cushioned by the good years, but newcomers, especially those on old-fashioned, out-dated holdings, faced disaster and either sold up or tried to modernise or diversify to meet public demand. The public never have much sympathy or understanding of producers' difficulties and do not make things easier. I well remember when we were suffering severe drought and lettuces were in short supply. I had a few at sixpence each, very dear in those days, and an indignant lady bitterly complained saying, "Any fool can grow a lettuce, sow a packet of seed in the ground and hundreds come up!"

Many growers were fortunate and sold out at the exceptionally high prices offered for land with planning approval and so reaped a far richer harvest than from tomatoes, fruit and other produce. Our land did not fall within that fortunate designation and besides we had taken to the life and enjoyed growing things.

Selling most of our produce in Yarmouth market meant that we were in competition with, at the time, a large number of very experienced market growers from the surrounding districts. They had the advantage of being experienced and well established and most had benefitted from the war years. Apart from a few odds and ends we offered neither quality nor quantity and had to sell cheaply to make a living.

When disaster struck we found our income drastically reduced and our plans for replanting and modernising the holding in jeopardy. We were still able to sell some fruit but most just fell off the trees and rotted. My wife Kathleen, who was a city girl, had never seen so much fruit as hung on our trees and was very upset to find that it could not be sold, especially when she squelched through our lovely Victoria plums, squashing them under foot. Sadly unsaleable.

As usual there were plenty of wiseacres telling us what to do. "Sell direct to hotels and boarding houses," they said and so I did. We loaded the van with trays of our best plums but only sold a few after a long frustrating day going from one hotel or boarding house to another. Even at half-a-crown a stone they were not wanted!

In the market things were just as bad, it was a buyers' market with too much on offer and imports taking the cream. Yarmouth was booming and people tended to treat themselves to oranges, bananas and peaches which they had missed during the war and the attractive foreign apples, pears and plums which were on sale just ahead of ours. Their incomes were increasing while ours fell and we struggled to survive. Unlike the farmers we had no feather bedding and many market gardeners justifiably decided to sell out and join the bandwaggon of prosperity but we decided we wanted to work with living things and looked to alternative uses for the land.

I must repeat, we were only small growers unable to employ as many helpers as was really necessary, always short of money and what we did have earmarked for tools, tractors, greenhouses or something else urgently required. If we had not liked the life in spite of all its ups and downs we could never have stuck it. With seldom a day off, no holidays with pay, we could not afford to be ill and were on the go from dawn to dark and even then often busy in the packing shed until bed time, willing slaves.

Another suggestion from the sidelines was, "Why not sell door to door? Old 'So and So' does it in Gorleston and is making a fortune." This was the late Jimmy Saunders, the so-called Mayor of Burgh Castle. I remember him well and his Trojan car, but he would not

4

have agreed about the fortune in spite of having a good business. Anyway, I decided to have a go. At that time Yarmouth was booming and fresh produce appreciated when delivered to the door. I had a bit of luck as a Mrs Lincoln was retiring and offered me her round. I had known her as a market gardener at Martham and for some reason she took to me and would accept nothing for her round which was generous. I repaid her by buying her old Morris lorry and many other items at her disposal sale.

The round was a large one and successful but posed difficulties with supply times and the weather. It was not easy to anticipate customers' needs and they expected us to carry everything, much of which we did not grow and had to buy wholesale. In addition to this we were going to market twice a week and the round soon became a burden to me. I was trying to serve two masters and it became obvious I must choose one or the other. The situation resolved itself as rapidly increasing competition turned into a cut throat business especially from the wholesalers. As my heart was set on growing I eventually gave it up except for a few of the larger boarding houses who gave me weekly orders and were good customers.

Another jolt came soon after. Several said, "Why don't you growers get together and form a growers co-operative?" Everyone is full of good advice for other people's business but it seemed a good idea and several leading growers formed what was known as E.G.M.A., the Eastern Growers Marketing Association and I joined, each member subscribing and then buying and selling through the association. It was fairly successful at first but lacked support. Through them we sold a huge crop of redcurrants, too big to sell locally, for a very good price. The well known Billy Beales ran the collection and delivery side and we supported it as much as possible but it came under the control of an unsuitable person and sadly it failed. I lost my investment but we did try.

Building up the business came first and we had to forego things other people took for granted. At Christmas my wife would ask, "What presents would you like?" It was a standing joke between us. I'd reply, "A new van, another greenhouse, another tractor." In what

5

little spare time we did have sailing on the Broads was our delight and we explored every inch gained up these narrow waterways with only sails and quant to rely on. In those days the water was unpolluted, motor cruisers not so obtrusive and holidaymakers better mannered. We could ill afford the time or money but it made a welcome break from the long hours and constant anxiety.

We were always handicapped by lack of money and had to earn it before we could expand or make necessary changes. At times we lived from hand to mouth and every investment was risky. A life assurance policy of many years matured and helped at one critical period and at another time my wife received an unexpected legacy which helped keep the wolf from the door. If my wife had not been so understanding, thrifty, hard working, supportive and steadfast, we would have given up after the early setbacks.

Always busy, my wife Kathleen is pictured here picking daffodils in one of our greenhouses.

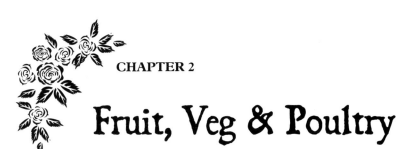

CHAPTER 2

Fruit, Veg & Poultry

An Old Home Swept Away
Gone alas in one fell sweep;
The poor old home is but a heap
Of rubble in a dismal pile,
With here and there a broken tile.
Old things, old folk seem in the way,
Soon pushed aside this careless day.

Russets and sweet Pearmain,
William pears and old Blenheim;
Dr. Harvey was our joy,
A mighty tree when still a boy,
Now just a pile of firewood,
A pyre of all that seemed so good

Gone are Prince and Prancer too,
Gone the sweet breathed Jersey cow.
The Guinea fowl up in the trees,
Ducks on the pond and hissing geese,
All alas have passed away
And modern fancies hold full sway.

In my opinion many fruits, vegetables and poultry do not taste as good as they did years ago. The scientific breeders have achieved quantity and uniformity but have lost quality and flavour. A good example is Cox's Orange Pippin, where they have produced an attractive apple to suit the supermarkets but have lost the unique

9

flavour which was world famous. The last time I enjoyed a real tomato was from a plant self sown growing in the ruins of the old greenhouses that collapsed in a gale. They had a real tomato tang to the tongue!

We all benefit from the wonderful work of our plant breeders and scientists, no-one should disparage their great achievements but we should ask them to pause, take stock of the needs of a changing situation and breed back any unique tastes, flavours and special qualities in danger of being lost forever. There is still time and preservation is most important.

Apple storage needs investigating. Years ago we stored our apples for months in very simple apple stores, the flavour was retained and in some cases improved by slow ripening. With proper precautions there was little trouble with vermin or frost. Today I believe they gas the poor things and it's hardly possible to distinguish one apple from another, there is a sort of uniformity of taste.

Our old orchard was very old, long since past its prime, a typical East Flegg orchard well protected by thick hedges of beech, blackthorn, Myrobalan (the cherry plum) and other hedging plants on or within a bank, a good windbreak. Inside was planted a further taller windbreak of the cultivated Bullace, not unlike a small gage plum. Under the Bullaces were drifts of the old English Pheasant Eye narcissus, beautifully scented. In the early days we were very busy harvesting our apples, pears, plums and other fruit.

The main section consisted of twenty-four Dr. Harvey apple trees, huge trees spaced well apart and needing tall ladders to gather the apples which hung from long twigs and were usually best at the top of the trees, a big crop of clean, disease-free fruit. The apples start green with small russeting on the shoulder. It is unusual in that it can have four variations on the same tree, even on the same branch, which is very confusing to the present day apple experts but is the hall mark of the true Dr. Harvey. One apple may be green, the next green with a russet indentation, another green with a pinky red flush on the shoulder but they are all the same old Dr. Harvey and will all turn a lovely golden yellow when ripe. They are one of the oldest

The apples we had in our old orchard were Beauty of Bath, Gladstone, Lady Sudeley, Irish Peach, Norfolk Royal, Allington Pippin, Worcester Pearmain, American Mother, Norfolk Russet, Ribston Pippin, Old English Pearmain, King Pippin, Codling, Warner's King, Lady Henniker, Charles Ross, Lord Derby, Annie Elizabeth, Newton Wonder, Blenheim Orange, Green Queen, Norfolk Beefing, Bramley Seedling and genuine Cox's Orange Pippins with the unique flavour sadly lacking in the apple of that name today.

The old trees were mostly large and tall when I was a boy, and every one was whitewashed making the trunks look like ghosts at night. In the spring the great mass of blossom billowed like a pink and white cloud while birds, large and small, flitted, sang and nested everywhere. It was Arcady indeed but now new houses stand on the site.

11

apples still surviving in East Anglia, as good as ever after over 350 years, and were always popular years ago for mincemeat and for laying away for Christmas.

Our early eating apples were Beauty of Bath, Gladstone and Lady Sudeley. Beauty of Bath was a difficult apple which if gathered too soon was rather sharp but if left until fully ripe fell off the trees at the slightest breeze. The Gladstone was a red-striped, soft sweet eating apple much favoured by old ladies, but now sadly, it seems to have gone out of cultivation. One of the best we grew was the old Blenheim Orange, a good-sized, long keeping and well flavoured eating apple. Another I greatly enjoyed was the Mother Apple or American Mother. It was a large flat apple for cooking or eating which always felt light in the hand, with a pinky red skin, juicy and refreshing to eat. I have not seen one for years.

A very rare apple was a small round russet, speckled green and white and among the sweetest I've ever tasted. Another we seem to have lost was a medium-sized, pointed, yellow eating apple, a summer eater and sweet. Very tasty and refreshing was Allington Pippin but it suffered from scab. We had only two trees of Cox's Orange Pippin, one about seventy years old but still fruiting well. It may have been from original stock, the apples were smallish, mixed red and russet but with the unique flavour of the true Cox. The other was about forty years old with slightly larger apples but also the real distinctive Cox's flavour. In those days the real test of a Cox ripeness was to shake the apple and if the pips rattled it was fit. Try that today, in vain. The scientific breeders have improved yield, colour, size and uniformity but lost the unique flavour. I've sampled Coxes from many shops and orchards but have yet to find a true Cox. Most are little better than the despised Golden Delicious.

An unusual old apple was Lady Henniker, usually long, ribbed and furry, hairy to the touch and long keeping for cooking or eating. Another eater was Ellisons Orange, a good cropper, earlier than Cox. Lady Sudeley was a fair-sized sweet eating apple, yellow with a red flush. We also had a red sport with the same flavour.

Warner's King was a big green cooking apple favoured for baking.

It had a reputation for scab but grew clean and well with us, one of the best. A great favourite locally was the great Green Queen, more widely known as Green Roland. In Rollesby it was said to have been discovered by a Mr. Shreeve. He may well have introduced it to Rollesby but it was known before his time. It is a large tall apple, green with a pink flush and it was not unusual for the fruit to weigh half a pound each. It was an eating, keeping, cooking apple with no acid and always free of disease and it kept until April. It is a regular heavy cropper without blemish. Why the nurserymen neglect it I cannot understand.

Among our younger trees was the Irish Peach, a good looking early eater but not very popular and Newton Wonder, a long-keeping cooking apple, very red and often enjoyed for juicy eating. Another long-keeping cooker was Annie Elizabeth. Our Bramley Seedlings were mostly younger trees too. It is unquestionably the best of all the cooking varieties and keeps so well. The Codling is earlier and much enjoyed by the few who really appreciate the differing qualities of English apples. Charles Ross is a splendid large eater but should not be kept too long while Lord Derby is a cooking or eating apple worth growing as is the quality apple Ribston Pippin, a fine eating apple

At first we had the Norfolk Beefing, a very old historic apple but could not sell it as people would not bother to keep it until fit or go to the trouble of preparing as of old. It was a delicacy long out of fashion. It was one of the longest keeping of all and enormous quantities were sold years ago. It is famous through the writings of Parson Woodforde and Charles Dickens but we grubbed it out as few would buy it.

We had a good selection of pears for cooking or eating of which Doyenne du Comice and Beurre Hardy were the best. When ripened correctly they became bags of juice which ran down your mouth. Old people used to describe them as a 'glass of wine,' a very apt description. Our William pears were old, good to eat but not as smooth skinned as the foreign imports and came later. An excellent early pear was Jargonelle, long like a Bartlett, very sweet and juicy but devils to gather as they attracted swarms of wasps which fell on our heads and necks when gathering. They came early and sold well. The Conference pears were lovely when ripe and I was fond of an eating pear known as Swarmer, they were so prolific! Sweet eating.

Before the art and pride in good cooking and prudent housekeeping went out of fashion stewing pears were in great demand and supplied a cheap, popular, wholesome sweet, especially with a bit of root ginger or cloves added. We had the huge Catillacs and Burgundy pears and sold hundredweights of them, the Burgundy especially although it was not a good keeper.

Our plums were another victim of foreign imports. We had great quantities of Early Rivers plums which we carefully sorted for eating and cooking and a huge crop of Victoria plums which I have never seen or eaten better anywhere, the branches often breaking under the weight. We also grew Czar, Pond Seedling, Black Prince, the Egg Plum and the Old Greengage, the most delightful greengage ever grown, far superior to the varieties grown today. The greengage was originally introduced to this country by Sir Thomas Gage who lived in Bury St Edmunds and they grow well in the East Anglian climate.

All around the north and east of the orchard, inside the strong hedge, grew the garden or cultivated Bullaces, not the wild or hedge Bullace. These trees were usually laden, the fruit selling readily for stewing, jam or bottling whilst the trees formed a useful windbreak.

We also grew many varieties of gooseberries, Golden Ball being a popular eating one. At first we only had old raspberries, Lloyd George and the Norfolk Giant, now almost unobtainable, also some loganberries and cultivated blackberries, figs and a splendid medlar tree with its unusual fruit looking a little like giant rose-hips. Some say that the flavour is like rotten pears and they are certainly an acquired taste, not enjoyed by everyone today but the tree is most attractive and not too large or tall. Some fine specimens can be seen at Somerleyton Hall gardens. We had only two red and white cherries but the birds got most of them. In our early day red-currants were a big crop, along with whitecurrants which were greatly enjoyed by old ladies with bread and butter in the days of afternoon tea, comfortable leisurely days.

We always appreciated crops that were little trouble to produce. Rhubarb grew easily and sold well; ours was the raspberry variety, red down the stalk and never coarse. Horseradish grew like a weed all over the garden and seemed a nuisance until we found there was a demand for it. In the heyday of market gardening it was cultivated in beds but escaped and spread everywhere. When we realised the demand we soon dug into this little gold mine. Taking long stalks we cleaned and washed them, tied them in bundles and displayed them on the stall where they sold like hot cakes at 1/- a bunch, a welcome addition to our takings.

Herbs were another small but useful line. We grew mint, sage, parsley and thyme in season. Mint was popular so we had a big bed but rust was a problem so we always had a young bed to replace it and parsley we often supplied to the six day stalls. In the early years we grew a lot of shallots, the brown English ones which we sold mainly to gardeners as seed, not allowed today I believe. They were much used for pickles and are far superior to the Dutch, greener, larger variety sold today. They were a handy sale in the spring. When dealing with other crops it was difficult to find time for it but somehow we managed.

In our early days we relied on a good start from early potatoes. This had always been a big help as part of the garden was sheltered, frost free and warm and grew early crops. At that time importers were not raking in early potatoes from all over the world and people willingly queued to buy and enjoy English new potatoes, the real McCoy, eagerly snapped up at a shilling a pound which seemed exorbitant for those days yet no one begrudged it for a long awaited treat.

Seedsmens' catalogues always fascinate me, they make lovely winter reading. After trying several firms we eventually chose Harrisons of Leicester, one of the biggest. Their representative, Mr Smith, so friendly and helpful, always helped us order for the coming season. Many new varieties of flowers and vegetables were appearing and as I was always looking for something different which our competitors did not offer I tried several. One early success was with a new type of Brussels sprout named Blue Jade which it was claimed would be ready in July. The seed cost five shillings an ounce, about five times the usual price, but being so early should easily recoup the cost and scoop the market. I tried it and it was a winner, a dark green sprout in July long before anyone else. It sold at 2/6d a pound, a very high price, and customers were clamouring for more. Our competitors were very put out but soon followed our lead. We grew it until it deteriorated because actually it was a bit of a nuisance as we had so much else to offer at that time. Sprouts are more useful with the winter vegetables.

For some time we had kept a dozen hens for home consumption, selling the surplus eggs on the stall. They were Buff Rocks, not the best of layers but large attractive birds, housed in a home-made house. A good friend, Billy Page of Scratby, urged us to go in for poultry as a useful sideline, rearing our own birds from day old chicks. On his advice we built a brooder house, ordered a Miller warm-floor brooder, feeders and drinking fountain from Boddy's of Norwich.

The brooder house was home-made but very roomy for a hundred chicks at a time. Just then there were American air bases all over Norfolk which were using large packing cases for aero-engines and other parts. An enterprising firm advertised the empties at a very low price so I ordered a load of these huge boxes and they were duly delivered. They had to be dismantled as they were far too big to handle but fortunately each one easily divided into six sections and I used five sections to make a brooder house along with two Dutch lights for windows and ventilation plus a second hand door. It was rough and ready but wind and rain proof and very solid. The remaining packing cases came in handy later.

The chicks were then ordered from Stirling, a well known firm, hatching day-old chicks. At that time Lord Fitzroy was one of the firm and very keen. The brooder heaters were lit, water fountains and feeding troughs filled and all was warm and cosy, ready for the

August 1955, Kathleen and me feeding the chickens.

chicks. What a thrill when they arrived, all those lovely golden balls of fluff, so bright eyed and lively! Gently we placed them in the brooder and in no time they found the food and water and were flapping their little wings as they ran in and out of their new home. It was amazing how quickly they adapted to these new surroundings. Like anxious parents we crept out at night to see how they were doing! Stirling always gave two extra chicks to allow for loss but we usually raised them all and poultry was a very useful and profitable side line.

All the chicken houses I built myself from second-hand wood from bombed houses in Yarmouth which was sold by auction, and some new timbers from Orfeur and Bellin plus new asbestos sheets and corrugated roofing and several second-hand doors. The windows and ventilation was from Dutch lights. They were large, light, dry and well ventilated with droppings boards and plenty of attractive nest boxes, also good deep litter to keep the birds active scratching for corn in bad weather. They had access to big grassy runs and were moved as the grass was eaten which was good for the hens and good for the land.

We had ample room for a hundred and fifty birds but usually kept around a hundred. Our point-of-lay pullets established a good reputation and we built up a regular demand, rearing extra when required, a profitable sideline for several years. Cockerels for sale were another useful line.

As we diversified things improved with our free range eggs all nice and brown soon attracting regular customers. We were amused by the preference for brown eggs as from experience we knew there is no difference between those and white, but the customer is always right.

The poultry industry had changed tremendously and the new hybrid birds were achieving very high egg production all the year round and it was goodbye to the familiar Rhode Island Reds, White Sussex, Leghorns and other old favourites. We spent much time with our birds from day old onwards and it paid off as they became tame and easy to handle. They were healthy and laid well but it was essential to keep them active or they developed bad habits such as feather picking which could lead to near-cannibalism. When we tried intensive egg production we suffered from that. We fitted 'hen specs' on persistent offenders, a sort of pince-nez on their beaks, blinkers which prevented them picking their neighbours. They were effective but I feel unnatural conditions are not good for man, bird or beast. We tried lighting at night for a time, but the poor things got all mixed up and made a dreadful commotion some nights and I feared our neighbours would complain.

There were always a number of wasps' nests around and young Graham Cooke helped dig them out. My main tool was a blow-lamp which I directed on to the nest's entrance. Wasps' nests only have one entrance and wasps entering or leaving had little chance of avoiding what to them was an invisible death ray - although a few did and had to be dealt with swiftly. The next move was to dig out the nest, keeping the blow-lamp handy to deal with any remaining wasps, and the grubs were given to the hens. Lovely grub!

Chickens use a lot of grit and oyster shell which we bought from George Edwards who with his wife had the popular shell fish and general stores in Howard Street, Yarmouth. He ground up all the empty shells and was pleased to sell them. Everyone knew George with his ruddy countenance, hair neatly plastered down, always smartly dressed and wearing his green baize apron. The shop had an attractive front with a delightful inner courtyard, all now demolished but all is not lost as Mr Malcolm Ferrow, the antique dealer, bought the shop front for future use.

Prices and sales of eggs fell and things became difficult for many small poultry keepers. We were not affected at first but later going through the accounts it was obvious that instead of the poultry keeping us we would soon be keeping them! Sadly we sold them at the end of the season. They had been part of our lives for so long.

CHAPTER 3

The Birds, the Bees & Other Animals

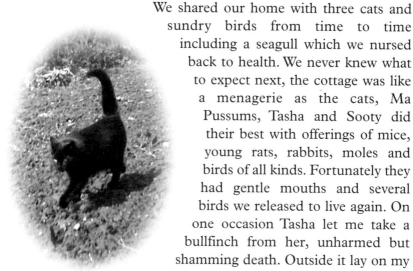

Sooty the cat

We shared our home with three cats and sundry birds from time to time including a seagull which we nursed back to health. We never knew what to expect next, the cottage was like a menagerie as the cats, Ma Pussums, Tasha and Sooty did their best with offerings of mice, young rats, rabbits, moles and birds of all kinds. Fortunately they had gentle mouths and several birds we released to live again. On one occasion Tasha let me take a bullfinch from her, unharmed but shamming death. Outside it lay on my hand, opened an eye and in a flash was gone. It was a beautiful bird but a terrible threat to our fruit buds: it is unbelievable the damage they cause to fruit growers in a few hours. I might have been justified in killing it but who could deliberately kill such a lovely creature?

Another time, hearing a terrible din we rushed out to find our cats guarding each end of a drain pipe lying on the path. They had trapped a weasel, a small but very fierce and dangerous animal, unafraid of man or beast. It would be dangerous leaving them to deal with such a vicious creature so I quickly got two syrup tins, put one in each end of the pipe, then carried it well away to a hedge, removed one tin and out flashed the weasel, hissed at me and was away!

Twice we rescued kestrels and what a job it is to feed those fierce little hawks. I made a special cage and we managed to nurse them back to health and freedom. In one case it was very touching as at the moment of release its mate swooped down and they both flew off displaying and mewing. Unbeknown to us they must have maintained contact during convalescence!

Sammy the seagull was a special case. He lived with us for fourteen years and was a great character. He was found with a broken wing and a shattered beak, so splintered he was unable to eat. The wing I was able to set but his constant efforts to escape upset the splint and the wing set at the wrong angle and he was unable to fly. At first I despaired of saving him as his beak was in such a dreadful state but the splintered fragments fell off and he soon grew a fine new bill and was able to feed himself which saved us much time and trouble.

Sammy the seagull

When found he was in a filthy condition and after feeding him with brown bread and milk we placed him in a basin to be washed. Weak as he was, no sooner did he feel the water than he struggled to his feet and had a tremendous splashing and flapping session. Imagine my wife's kitchen! We knew then he was going to live and every day he grew stronger. For safety we kept him in a shed but as he recovered he had the run of the garden which we made secure against dogs etc. We sank an old dustbin lid in the ground and he loved to bathe in it and we had to keep re-filling it with water. Later I made a bigger pool from a large sink. The soil dug out made a big heap and after bathing he would mount the heap and tread the soil with continuous stamping movements and in springtime, the breeding season, he nearly drove us frantic with his raucous calls, especially when screaming at the kitchen door.

He was a blackheaded gull and like other gulls ate anything. After he recovered we offered him some sprats. He seized a large one, threw it up in the air, caught it head first and gulped it down. The bill is not just a V-shape but opens up like a square to accommodate large objects. It was fascinating to watch the sprat slowly disappearing.

The funniest sight was when Tasha came down the path carrying a live mouse in her mouth. In an instant Sammy snatched the mouse, still alive, and the last we saw was the tail wriggling before it disappeared. We never regarded Sammy as a pet and would gladly have released him but for his gammy wing.

My happiest experience was when a dog forced its way into the garden and rushed at Sammy who ran shrieking and crouched between my feet for protection. I was much moved to have a wild creature run to me for safety. The dog belonged to Jack Bradley, the then landlord of the *Jolly Farmers* at Ormesby, a former well-known footballer. The dog repeated the performance a few days later and actually caught Sammy in his mouth but being a gun dog did no harm and Jack released him unhurt.

Sammy always followed us round when we were weeding or tidying the cottage garden, picking up earwigs, centipedes, woodlice, small worms and any tasty morsels. As he became old he lived mostly in the heated greenhouse. In the summer he lost his black cap and moulted but in spring he resumed his splendid black feathers, his bill was a brilliant orange, also his legs, and his eyes gleamed. It was his breeding plumage and he looked fine. At one time he became very ill and we feared we would lose him but he suddenly recovered and lived a few more years, finally dying after fourteen years.

At one time we had another blackheaded gull which we revived and put in an empty greenhouse with Sammy for company but they completely ignored each other. I was most impressed when it began flying up and down in the restricted space, 50 ft x 12 ft. Many birds fly against the glass but his was a controlled flight, carefully looking for a way out. When released on Ormesby Broad he flew then landed on the water to enjoy great washing and splashing, as if to say let me wash those human hands off me!

In our early years the garden and orchard was home and refuge to many birds. The great old trees gave shelter and nesting places to so many, large and small, and the banks and hedges just as many. The dawn chorus was overpowering. Woodpeckers worked the old trees, yaffle, great and lesser spotted and the active little tree creepers, constantly exploring every trunk.

The twittering and tinkling long tailed tits with all their cousins merrily worked their way along the hedges and away to Scratby Hall woods and back. There were robins in the banks, and wrens and hedge-sparrows in all the hedges. Every year the shy woodcock visited us in spring and pheasants and partridges foraged through the crops. Cock pheasants often fed with the chickens and on one occasion two cocks were so exhausted in a fight I was able to catch one. What a magnificent bird to hold in the hand! I was so impressed I took it to show Kathleen. We might easily have had it for dinner but how could one deliberately kill such an attractive creature?

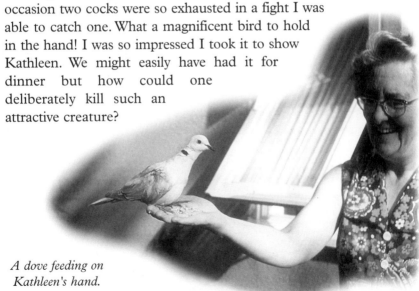

A dove feeding on Kathleen's hand.

Every year while the tree survived a pair of mistle thrushes nested in our tallest pear tree and reared their young. Hedgehogs hibernated in the hedge bottoms and we even harboured the dainty harvest mouse. Toads and frogs flourished. There was a huge toad that lived under a water butt and in our propagating house we encouraged a family of toads, father toad, mother toad and four tiny totty toads!

When ploughing I'd spot some, stop and move them to safety, or I might turn up a nest of that 'wee, sleekit, cow'rin', tim'rous beastie' on a furrow. At times clouds of seagulls would follow the plough, so eagerly snatching the worms as to get ploughed in and I had to stop and extricate them. Among our rarer birds we occasionally saw redstarts and black redstarts. For many years cuckoos were calling in all directions and the poor little hedgesparrows suffered. I was good at mimicking the cuckoo and many an indignant bird mobbed me thinking I was a rival but by 1978 we rarely heard one.

In 1947 I remember hearing the last corncrake in Ormesby. Snipe used to drum across the sky in dozens while swifts, swallows and martins were swooping everywhere. Owls and hawks of every kind were plentiful and every winter skeins of geese passed over in hundreds. We had huge flocks of fieldfares and redwings, starving hungry, massed in the orchard desperate for any windfalls or apples we threw out. I took an exhausted redwing indoors once and was appalled by the multitude of mites that descended from it. Like most thrushes they were hard to save.

One dainty visitor was the waxwing which feasted on berries under our windows. Other attractive visitors were linnets, blackcaps, firecrests, garden warblers, goldcrests and wagtails while overhead the skylarks sang joyfully. Sadly by 1980 bird life in Ormesby was greatly diminished.

I had a tame robin who loved poultry pellets and even took them from my mouth. It is dangerous for wild creatures to get too tame as they lose their natural caution, often with disastrous results. At one time we had a very friendly collared dove which lighted on our heads or shoulders when called, rather disconcerting if unexpected, and always ate from our hands. Unbelievably friendly and always greeting us as we returned home, one day it was missing and we never saw it again. I feel sure it was too friendly and someone caught it and kept it captive as a pet.

For years blackbirds ran in and out of our kitchen and the cats observed an uneasy truce. Opposite our windows was a huge King Russet apple tree. The apple had developed bitter pit so we grew variegated ivies up it as a show piece. So many birds used it that we called it 'Bird Hotel'. Unwelcome guests were the pigeons and house sparrows. One day we startled a jay swooping on some sparrows, literally tearing one to pieces before our eyes.

Many birds meet in our garden,
Coming at the break of dawn,
Flitting, hopping, chirping, singing,
Some displaying on the lawn.

Cheeky sparrows in their dozens,
Robin jealous and alone,
Great tits, blue tits and their cousins
And a yaffle on his own.

There's the blackbird sleek and shiny,
Mark his shiny yellow bill,
Hard worked when the brood is tiny,
All those gaping mouths to fill.

One more nervous than the others,
See that lovely spotted breast.
Many snails he has uncovered
To crack on the anvil with the rest.

Does he look or does he listen?
Watch the thrush upon the lawn;
How that beady eye does glisten,
Seeking out the slimy worm.

Although his song the rest might envy
He is for all a timid bird,
Last to take the offered titbit,
First away, looks so forlorn.

More assured, the lonely stormcock
Comes with winter's cruel blast.
Chance time comes the dainty waxwing
Modestly he bears his crest.

But those that serenade the morning
Are the ones I like the best.

✳   ✳   ✳   ✳   ✳

We also kept bees. I really took to them, and we used the honey in a variety of ways, my wife using a great deal in cooking, especially in her cakes. My mother had always wanted to keep bees and during the war bought two Taylor hives and tried to find some bees. Unfortunately the nearest beekeeper did not approve of female beekeepers and found excuses for not having any swarms! When I returned from the war and was settling on the land Mother gave me the hives and I approached our local beekeeper for advice. He was most enthusiastic and told me where to set the hives and promised me his next two swarms. On his advice I then purchased a bee veil, gloves, smoker, feeders and a nasty looking knife for the combs.

Stanley Whitby was a real expert and most encouraging. He soon settled two good swarms in my hives and showed me what to do, helped me find a second-hand extractor and the other necessary equipment. When the hives were full of honey, he showed me how to take it, extract and ripen it and he could not have been more patient and helpful and would take nothing for all his trouble. Another good friend.

Do not keep bees if you are afraid of them! They never worried me even when they were angry and aggressive and I was seldom stung. However, on one occasion I was over confident and careless. I was examining the frames while the bees seemed happy and quiet, so quiet I neglected to wear gloves or make sure my veil was securely tucked into my jacket. Suddenly the bees became angry and attacked me viciously and got in under my veil. The hives were open, the frame in my hands so I quietly but surely put back the frames and reassembled the hive. By this time the bees were in my shirt and everywhere and were beginning to sting! Rushing home I shouted, "Shut the door and windows!" as I appeared in a cloud of bees. My wife saw my predicament, shut the bees out and awaited developments. Running fast up the lawn I eluded the bulk of the bees, opened the scullery door, rushed in, yelled to my wife to shut the kitchen door and then divested myself of hat, veil, and coat, opened the door and shooed the bees out, shut the door, took off my shirt and vest, shook out most of the rest of the bees and got them outside somehow or other. I was still crawling with bees which were determined to sting me and unhappily had to be killed. I was never so careless again.

We usually had a good supply of honey until agricultural spraying almost destroyed our bees. Once all the bees died and we blamed spraying. The hives stood empty so I cleaned them up, ready to try again. While I was hoeing one day I heard a noise in the sky and to my astonishment it was a huge swarm of bees. In no time they swirled across the sky and I followed them into our next field where, to my amazement, they swiftly descended and lit upon an empty hive. Their queen quickly found her way in and, in no time, the rest followed, a welcome sight. Unbelievably another swarm later made their home in our second hive. They were both good swarms and provided us with honey until we finally sold the lot.

Bees swarm and rest in the most extraordinary places and are not always easy to collect. It's wonderful to see them hanging glistening like satin, constantly in motion. Often it is straightforward and easy taking the swarm, and after a rest watching them streaming up a slope into the hive. It is interesting to watch for the queen making her way through the workers and into her new home. If everything is satisfactory they soon settle in and start filling the combs with honey but if they do not like the hive they come out as quickly as they went in. Old Parson Woodforde had that experience all those years ago.

CHAPTER 4

Flowers – the Odd Bunch or Two

In 1950 foreign imports swept the board and English produce took second place, only the best of local fruit holding its own. Many English growers cut their own throats by flooding the markets with unripe, ungraded, inferior samples which gave English apples a bad name. Fortunately there were notable exceptions and they still survive. I have always believed there is nothing better than English fruit gathered and presented at the proper time. Had I owned all the land we cultivated it would have been replanted with the best varieties and also a selection of hollies and foliage plants as they sold well especially at Christmas. Holly trees are a long-term investment as they take so long to grow. After a few years if carefully trimmed they yield a good return for little trouble. We developed a very good trade for bunches of mixed greenery for Christmas which gave a useful income at that period.

About the same time as we moved into poultry we made another investment in an attempt to replace our falling fruit sales. My brother Pat knew a neighbour who had had enough of market gardening and was selling out. He only wanted £60 for an excellent 50 ft x 12 ft Dutch-light greenhouse complete with base but it needed dismantling, carting home and re-erecting. He also had a similar structure, dismantled, lacking a base, but with all the glass and some extra, all for £30. Among other useful items were a large number of galvanised ring pots, a real bargain which we used until we retired and then sold for a good price. There was also a large selection of herbaceous plants which sold well later. The only drawback was that they were riddled with shamrock, a pernicious weed which spread rapidly, a pest. All this was a good bargain but took every penny we possessed.

My brother and I went to Beccles and dismantled everything ready for collecting. I had arranged with Redvers Lawson, our local coal merchant, for one of his lorries and a driver to collect and deliver to Ormesby so all was successfully accomplished.

How I found time to clean, paint and erect this new addition I do not know as two days a week were spent at market and another two days preparing the produce. Previously we only had two dilapidated old greenhouses, beyond repair and liable to fall down in any high wind which they did shortly afterwards.

Fortunately the new addition was erected in time for a crop of tomatoes and as we had no propagating house we bought the plants from another good friend, Ernie Morse of Belton. They were Stonor's Moneymaker, well named as we had a tremendous crop. Being on fresh ground helped and they sold out every week. Gathering and working with tomatoes was not as nice as it looks. There is an awful stain from fruit and plants and is hard on the hands. I often used gloves but still had to do a lot of handwashing.

It was another venture that helped put us on our feet financially. Following the tomatoes we planted latish chrysanthemums but that was a chancy crop as we had no heating. Later I bought commercial Aladdin heaters to keep the air moving and frosts out, aided by polythene sheeting. It was risky but we were lucky and had profitable crops. More Dutch-lights were used on frames to bring on seedlings and later to protect chrysanthemums in a temporary structure. It was all improvisation to save expense.

Like most growers we had marigolds, statice, asters, cornflowers and many other cut flowers which were over produced and difficult to sell unless sent away to London, Liverpool or Newcastle, a costly and uncertain business. There was, in those days, a good rail service to the Midlands and Liverpool and we sent chrysanthemums to Hardy's at Newcastle and Fitzpatrick's at Liverpool, then the cost escalated and we sent no more. Yarmouth became our real market.

We had inherited a bed of excellent Parma violets and although fiddling to tie, they sold well. We also had some old fashioned cottage pinks but there were too many around then and we gave up growing them, which I now regret as there are none about and they had a wonderful scent. We had quite a run with lily-of-the-valley but they faded out for no reason at all.

Our largest and most successful crop was early chrysanthemums. At the wonderful shows in St. Andrews and Blackfriars Halls, Norwich, and at Yarmouth Town Hall, horticultural societies and leading firms displayed an outstanding selection of chrysanthemums. Famous names such as Woolmans and Stuart Ogg and others competed to attract amateur and commercial growers to their newest introductions or established favourites. At one time we tried dahlias but it was too much to manage them with chrysanthemums, they were so prolific.

In the members section there was tremendous competition and two of our neighbours contended for the highest awards. Bishop had the Hemsby family holiday camp and his gardener Walter Smith shared his enthusiasm. They produced some wonderful exhibits and there was great rivalry between them and Tyson who was a farmer.

We were very impressed and came home laden with catalogues but it took time to discover the best for market trade and we chopped and changed before we found what we wanted. At one time we bought 20,000 plants from those fine growers Herbert Gayton Ltd of Bradwell but they ceased supplying the trade and we turned elsewhere. Plants coming a distance suffered from overheating and were often unuseable so I arranged to collect from the nurseries.

When we changed to Challis of Poppleton near York I travelled 400 miles there and back but it was an opportunity to meet the supplier and discuss varieties. They deducted the carriage which offset the petrol. I did the same with other firms but we finally changed to Frank Rowe who we kept with until we finished. Chrysanthemums demand a great deal of attention from start to finish.

After securing satisfactory stock we took cuttings from them for the next season which we struck in boxes then planted out. With some hardy, very prolific varieties we used the lazy man's way and pulled off Irishman's cuttings, i.e. with the root attached. With some sorts this is quite satisfactory but you must make sure the stock is healthy.

In the early days each plant had to be staked and tied. We used bamboo canes which even when new snapped off in high winds causing extra work or damage if the flowers were in bloom. We grew mostly sprays at first but as our experience increased we bowed to customer demands and produced more medium-sized blooms.

The NFU horticultural secretary at the time suggested we increase these as on his own nursery he grew over 20,000 blooms and never had enough. Later I was glad we did not do as he suggested as there was a terrific gale and he lost most of his crop. The sprays suffer damage but are seldom a complete loss. Blooms are risky as they tend to rub or shatter and need careful handling. After several years of using rolls of wire to support plants netting squares became available and made things easier as they only needed lifting as the plants grew. They held them firm and prevented rubbing. Disbudding needed care and took time but gave a lovely assortment of blooms. Eventually we had an excellent assortment with good colours and reckoned to plant half new and half of our own growing.

We offered a selection of colours or mixed bunches of carefully selected bunches to avoid delays as customers tried to match colours while a queue impatiently waited to be served. We had a busy market stall with a reputation for good quality, reasonably priced chrysanthemums and we were rushed off our feet and could not afford to wait while customers hummed and hawed and could not make up their minds. We had to be firm but tactful or they would wreck other bunches. Blooms were worse if the customer could reach them. We had taken great care to offer newly-cut blooms that would last well only to see impatient ladies snatch their choice and wreck two or three others in their haste so after a while we selected the blooms or bunches at their direction. Much of our trade was in weekly orders from regular customers who trusted us to make the selection and that worked well.

One good customer was the cemetery and in later years we built up a big trade with local florists and we were stretched between shops and stall. We were lucky when we started to expand our chrysanthemum trade as several growers were retiring at the time and we happily filled the gap. Customers often said how much they would like to see the growing flowers and could not understand that there was little colour to be seen as we cut so closely to give them flowers that lasted well. Cacti buyers were the worst and could not see how the interruptions upset our routine and wasted so much precious time. Some were offended when we refused to entertain them. Trade customers were the only welcome visitors as they knew what they wanted and, like us, had no time to waste.

Among the thousands of chrysanthemum plants there were many short stems that grew into fine uncut flowers and it seemed a shame to see them wasted so I gathered some and tied short bunches. To my surprise they were snapped up. They met the need for some people and we gathered more and more to meet the demand. From then onwards we gathered all the short stemmed flowers and sold hundreds of bunches each week at 6d, 8d, 1/- and 1/6. It made a lot of extra work as we only attended to them after the main crop had been gathered and then tied them after tying the others and we were tired, but it was a profitable sideline well worth the trouble with flowers otherwise wasted. The prices seem trivial by today's values but were good for those days.

When we first grew chrysants we followed the spraying programme suggested by the advisory service to control capsid bugs, leaf-miner and other insects. One day I noticed a number of small birds methodically moving from plant to plant, picking off insects, and I wondered how the spraying might affect them. If they destroyed the pests, why spray? From then on I ceased spraying and, apart from very minor damage, our flowers were as good as any others. It may not apply in every area but we had the birds and they did the job. In the greenhouses where the birds could not operate we used smoke canisters. Birds are not always helpful. I've already mentioned bullfinches and any gardener knows the damage that can be inflicted by sparrows. We were near Trett's Provender Mill which attracted thousands of these pests. At one time they settled upon our newly planted chrysants and broke them right off and so serious was the damage that we feared losing all the plants. In desperation we put up twigs with black cotton stretched across the rows. It discouraged them but needed constant renewal. The problem was finally solved when we introduced wire mesh supports which we lay on the ground at planting time, planted in the squares, then moved them upwards as the plants grew. For some reason the sparrows disliked the wire and we had no further trouble. They turned their attention to polyanthus and other plants out of spite!

Gathering flowers seems lovely work to those who only gather the odd bunch or two but in actual fact it is hard and tiring work, walking up and down the the rows selecting, cutting, carrying increasingly heavy armfuls to the packing shed, and if raining with water running down sleeves and neck.

Much care had to be taken to detect and destroy any diseased plants on the land and in the greenhouses as trouble will soon spread. Good hygiene is essential in glasshouses and we carefully watched our chrysanthemums and ruthlessly destroyed any diseased plants. After we had built up a healthy saleable stock of chrysanthemums I made a silly mistake. A friend had a stock of older varieties with colours useful for mixed bunches and he let me have some. Unbeknown to him or me they were infected with a virus which soon spread to our previously clean stock. We immediately destroyed all the infected plants including some of our best varieties. It was heartbreaking but had to be done thoroughly. We replaced with new clean stock and refrained from growing chrysants on the infected land for two or three years. It was a setback but we soon recovered.

Christmas was a profitable time for flowers but chancey with cold houses. One year I was a bit too clever and planted the biggest house with a selection of Mayfords, mostly disbudded, and lost the lot! They did well and were about to burst into flower when we suffered a terrible frost for two or three nights which penetrated the house disastrously and we did not gather a single flower. I learned from my folly and selected more suitable types, introduced portable heaters, lined the house with polythene and never got caught again. In the future we grew Galaxy and Elegance in mixed colours and had good success.

We tried several sorts of cut flowers with varied success. One name amused us, 'Yunanensis Napsbury'. This had attractive blue flowers but was a poor seller. Pyrethrum was one of the best sellers but was overdone in the market. We had great success with 'Esther Read' and 'Horace Read' as we were among the first to have them on Yarmouth market. They sold well to the public and florists. 'Esther Read' was like a white daisy and Horace had a big reflex flower but a weak neck. They were terrific sellers for a time until everyone grew them. Outdoor freesias were also very popular and had a lovely scent.

At one time we grew a good selection of outdoor tulips including the parrot tulips and we did very well until Jack Howlett started bringing in the tail end of the Wisbech and Spalding forced bulbs which we could not compete with. This did not worry us unduly as we were switching to cacti which were far more profitable. I made special trays to carry them in the van, non-spillable.

Pot plants, cacti and succulents suited us well. If unsold they kept growing and were often more saleable after several weeks so there was little waste as with perishables. We avoided the larger and more delicate plants as they were difficult to cart, occupying too much room.

To pot the ever-growing number of plants we needed a lot of compost, sterilised compost. I knew all about John Innes mixtures having attended courses at Burlingham Horticultural Centre and thought if we had our own steriliser we could provide all we needed so I bought one. It proved to be too small and a failure as the ingredients were not easily available, and in any case I lacked time to do it properly. I had to admit failure and turned to the horticultural suppliers for cacti mixture, potting composts, grow bags and horticultural sundries. We also used farmyard manure and mustard which I ploughed in as I felt it improved the texture of the soil. All our sprout stalks and spent greens I chopped up with our big rotavator and ploughed in deep. When we cleared out the poultry droppings and deep litter it was very good for greens. Carefully rotating the crops kept us free from disease and other troubles.

We tried to concentrate on things we grew well. It pays to grow a small area well, rather than attempt too much not so well. It became obvious we needed brain and brawn for success and a little extra thought saved a lot of trouble. I frequently had to use my previous business experience to stand back and see how we were doing and what change of direction was needed.

CHAPTER 5

Diversify to Survive

It may sound as though all our efforts met with success but as everyone knows, nature deals some nasty shocks. Strawberries are generally a profitable crop and we had a small bed of that old favourite Sovereign, one of the finest strawberries ever grown, but a bad traveller. It had to be sold fresh and picked soon. We decided to plant a very big bed of the best modern variety and consulted the Horticultural Advisory Service. A Mr Bamford, I believe, came and approved the site, gave good advice and recommended a good certified stock. He was most helpful. I prepared the land as advised and watched the plants grow promisingly. We did not expect more than a berry or two and certainly did not anticipate drought. Our land was always difficult in dry seasons and that year the plants just withered away and we did not see a single berry and seeing no sign of recovery I ploughed the lot in that autumn.

When blackcurrants were booming we were tempted into joining the gold rush. John Trett, the miller, had replanted his raspberry land with blackcurrants and urged me to do the same as the price was 2/6 per pound, very high in those days, and Ribena were taking most of the crop. I followed his advice and set out a good block of new bushes but sadly by the time they were fruiting the price and demand had tumbled and all we could get was 6d per lb or 9d at most. We had a lot left unsold so Kathleen made dozens of jars of jam, bottled as many and extracted pints of lovely pure blackcurrant juice. It was all very disappointing but we must have been bursting with vitamin C! We were often short of money but never lacked good healthy home grown produce.

I became interested in growing cacti as a result of reading about a Mr Massey of Spalding who had built up a big business in cacti and succulents and supplied the trade. I went to see him and purchased a large selection and made arrangements to collect further supplies. All were named and labelled ready for sale. It was very much a shot in the dark but was unbelievably successful and they sold like hot cakes! It was the beginning of the cactus craze and we sold hundreds of these fascinating plants. We had no greenhouse available at the time but a neighbour allowed me to use his empty greenhouse for a while. However, we could not continue using this and, as we needed a propagating house, I urgently contacted Arthur Farman, our local carpenter and asked him to build us a 40 ft lean-to on a south-west wall. The brickwork was done by Bob Leath, the paintwork and glazing by Billy Huggins and the electric wiring and heating, including a very efficient thermostat, by Frank Chubbock, all of Ormesby, all carried out quickly and well.

This again took all our ready money so I timidly approached our bank to beg for an overdraft to see us through to the next year. I had only once overspent and had received a sharp note drawing my attention to my account being £20 overdrawn, hoping that I would soon correct this! I entered the bank feeling like Oliver Twist but the assistant manager listened to my tale and, to my surprise, agreed to a sum that would easily see us through until the money came rolling in. In all my dealings with bank and inland revenue officers I always told the truth and underestimated my expectations, fulfilled my promises and built up a relationship of confidence and trust which I valued.

There were many setbacks but the cactus craze saved our bacon. To meet the increasing demand we fitted new extra shelving and a huge heated bench in our lean-to greenhouse and set about meeting the amazing interest and demand for cacti and succulents. I again visited Mr Massey at Spalding and arranged for an even bigger supply. The public really went mad over these unusual plants and it certainly put us on our feet financially. Massey's were unable to cope with our steadily increasing demand as they had so many other

To meet the cactus demand we had to fit a huge heated bench in our lean-to greenhouse. Both pictures show Kathleen and me working among these fantastic plants in 1963.

customers so I found a supplier in Holland, and occasionally bought from America.

We received an amazing selection in bulk, all of which had to be potted and grown on to selling size and condition. There was always difficulty in getting enough so I began growing my own and, as constant high temperature is essential, I bought a large propagating frame controlled by an efficient thermostat to maintain 70°f. at all times. I also made a special large, under-floor heated propagating bench where it was easy to strike cactus and succulent cuttings and this quickly increased our supply of these ever-popular plants.

Our financial position was transformed so we erected another Dutch-light greenhouse on a brick base, built by George Palmer. In these small horticultural enterprises it is not wise to keep all your eggs in one basket and I had a feeling the cactus craze might subside, so we expanded our poultry side, increased our outdoor and indoor flowering chrysanthemums and started to grow early and outdoor bulbs. We were fortunate in meeting Mr Delf of Yarmouth who was the agent for Van Der Zalm of Holland, the well-known Dutch bulb growers who supplied us for many years.

By this time we had one full-time helper and needed more as horticulture is very labour intensive but it was almost impossible to attract really experienced workers as wages and conditions were so much more attractive in the holiday camp and building industries. The routine with the two greenhouses was bulbs for forcing, then tomatoes, followed by late chrysanthemums while the lean-to was kept for propagating and the cacti. We still had a lot of fruit but no longer relied on it.

In 1956 our neighbour, Major Miller, asked if I would like to take over his raspberry and apple enterprise. He was a retired army officer who had planted Malling Promise raspberries and Cox's Orange Pippin. He felt unable to continue and seemed to have a share cropping arrangement in mind, but I explained I was only interested in acquiring his enterprise lock, stock and barrel, subject to valuation as was usual in such cases. He was a reasonable and likeable man and we had George Edrich of Acle make the valuation and agreed to accept his figures. It was less than Major Miller expected and more than I expected to pay, so likely it was about right. Most people thought I'd bought the land but it was actually owned by relatives of Mrs Rolfe just opposite.

I soon started work on the land, grubbing out the Coxes as they are no use near the coast, more trouble than they are worth. In their place we had more poultry in another home-made chicken house, and cockerels in arks. These we sold to Dick Perry of Burgh Castle who paid a fair price and collected, and he also bought our old hens.

Raspberries were often a problem. We accepted many pickers but they did not always turn up if better jobs became available. In wet weather hardly any came and we regulars had to cope as best we could. There were several very wet seasons, so much so that Tony, our full-time helper, jokingly demanded danger money as he said that he was in fear of drowning, there was so much water on the rows. It was a dreadful job as we often had to change clothes three times a day. Sutton's, the canners, supplied the 2 lb baskets and collected twice a day. In addition we sold hundreds of 1 lb punnets in the market but the canners took the bulk of the crop. We had a few pick-your-own customers but did not encourage it. One of our PYO regulars was Mrs Clayton, well remembered as the able and popular landlady of the *Avenues Hotel* in Yarmouth and later as owner of a guest house in Ormesby. Among her many guests was a team of Turks learning weather lore at Hemsby weather station. They were conscripted to gather raspberries for her and knowing our difficulties with pickers she persuaded them to pick for us. Mrs Clayton was such a formidable and remarkable lady (with a heart of gold) they dare not refuse and were a great help. They refused any payment but as they liked honey we gave them a huge jar which they greatly enjoyed. Only one spoke English and what our pickers thought of them I cannot say, but they certainly liked the ladies!

We extended the canes with Malling Jewel and had a few good seasons but the difficulties with pickers decided us to finish with raspberries. We weighed things up and decided it was time for a change. I had been considering increasing our greenhouses and was attracted by the mobile houses of Robinson's of Winchester. Their enthusiastic and helpful representative called, and we decided to buy a 50 ft x 20 ft mobile Dutch-light structure with steel tracks to cover three sites. By then we were in much more favourable financial circumstances but it was not convenient to pay cash and as I do not like hire purchase once again I visited the bank. I was fortunate enough to deal with the manager, Mr Rivett. He was most helpful and extended my overdraft enough to frighten me. He seemed impressed with our steady progress and the fact that we always achieved our objectives, but banks are always cautious and at his suggestion, actually it was really more than a suggestion, I took out a substantial life assurance policy and as a firm believer in life assurance doubled it as extra security.

At the end of the season I grubbed out all the raspberries, ploughed the land and arranged for delivery of the new greenhouse. It arrived - steelwork, wooden sections, steel track, concrete piers, sliding doors, ventilating gear etc, etc, and crates of glass. I had invited our nephews, Kyle and Desmond, to devote some of their holiday to assist with erecting what seemed to us a huge greenhouse. We quickly unloaded the lorry, followed the instructions, set the base out square, laid the rails and assembled the steel framework on wheels. We were thrilled to see a large greenhouse rising from bits and pieces. Our nephews were still schoolboys but enjoyed every minute of seeing it grow before our eyes. The glazing was the tricky part as handling such big sheets of glass even in the slightest wind was risky. At last we fixed the ventilating gear, fitted the plastic skirting, hung the sliding doors and all was complete. It was only a small greenhouse really but to us it looked vast and we were bucked to think we had built it.

To test mobility we extended the track and pushed the house backward and forward. It moved easily but was beyond my single-handed efforts. On my own I used the tractor with long nylon ropes to draw it over the sites. The ends hinged upwards and allowed movement over late chrysanthemums, the last crop to be covered. I had water laid on, stop cock, meter and hose connection. This also

My nephews, Desmond and Kyle Holmes, assembling the framework of our new greenhouse in 1967.

The greenhouse rises from all the bits and pieces as Desmond puts some finishing touches to the roof.

47

came in handy for early outdoor chrysanthemums using sprinklers and a perforated hose. In those days water was cheap and plentiful.

Traditional growers did not favour mobile glasshouses but they suited my needs. Firstly, they were much cheaper than ordinary greenhouses and I could erect them on my own if necessary. Secondly, they could be dismantled and easily re-erected on another site which was particularly important to me as I only rented much of the land and it would be foolish to erect permanent structures with possible complications on quitting. Thirdly, mobiles avoided soil sickness prevalent in fixed houses growing the same crops every year with all the sterilising problems involved.

Having three sites worked well. I planted Delta lettuce and anemones, then I drew the house over site two and, when the soil warmed, planted tomatoes by April 23rd. We only grew cold house tomatoes but the glass came within inches of the soil, gaining maximum heat from the sun and the plants grew away quickly. Having our own propagating house meant we had the plants all ready.

We now had three growing houses and I planted the big one with a Dutch hybrid which gave a heavy crop of good quality tomatoes, the second house was filled with Moneymaker, a popular reliable cropper, and the third house was half Alicante and half Big Boy, a remarkable American introduction, huge but of excellent quality, good for slicing and very popular once tried.

The holiday trade was booming and local tomatoes were in great demand. Visitors returning home demanded them extra firm and under-ripe so we made a point of catering for them which helped us as we cleared the crop twice a week and we seldom had over-ripe tomatoes. In the top weeks I gathered $\frac{1}{4}$ ton twice a week which shows how good the demand was in our day, and we were only small growers. Small tomatoes sold well and also the extra large ones. Tomatoes were our main crop in July and August, followed in later years by chrysanthemums until the second week in January and then we fell back on vegetables and apples and a fairly thin time until we got busy again in the spring. Actually, we were not sorry to ease up a bit.

48

Ever ready to diversify, we also tried ornamental gourds, grown like trailing marrows and fascinating with their strange shapes and attractive colours. The Turk's Head was one. We grew a large mixed assortment to the delight of the many flower arrangers so active at that time. Gourds had to be naturally ripened and kept well, even better if varnished. A similar attraction was the Custard Pie marrow which was creamy white and could actually be eaten but was in great demand as a floral decoration. With the gourds they flourished until wet and frost finished them off. We sold hundreds and I even showed them on television!

Another surprising success was with outside cucumbers and gherkins. I was very intrigued by one called Burpee Hybrid. Cucumbers are said to make you burp so what about this one! It was an American introduction by a Mr Burpee and was described as having the flavour and quality of the best greenhouse cucumbers but easily produced outside. Once people tried it it was a terrific success and we gathered hundreds. They were short, fat and stumpy but had a delightful flavour, slicing beautifully with no faults. We also grew gherkins for pickling.

One customer regularly bought a large quantity of gherkins. She and her husband, a sea captain, had fled from Latvia and had been living in England ever since. They pickled and used gherkins and outside cucumbers as in their homeland and also sold them in jars to their compatriots and neighbours. Other customers were Indian, Italian and other Europeans who made great use of them. Many Indians took them back to London, Bedford and other places. It was an interesting contact with other peoples and customs.

Courgettes were becoming popular so we tried them. One was most prolific and soon grew into a fine large marrow which sold better than the courgettes so we concentrated on marrows. Vegetables have a cycle of popularity and our growing marrows and gherkins coincided with such a cycle. We grew hundreds and I worried about my van's springs, the marrows were so heavy!

We tried pumpkins but they were too bulky. Norfolk people refer to them as 'millions' as there are so many seeds. Million pie was a great favourite years ago prepared as a savoury or a sweet.

I had expected the cacti boom to burst and was preparing to extend our range of plants when we were forestalled by the weather. In 1962 we suffered one of the worst winters I can remember. There was no snow to insulate the greenhouse roof and an intensely cold wind penetrated everywhere. We had our cacti house full of young plants growing well, part of our stock for the next three years. Until then our heating had been adequate but despite all our efforts the temperature fell disastrously. The result was not immediately apparent but in the following weeks we threw away hundreds of frosted plants. It was a setback but we grew geraniums, coleus and similar plants which sold quickly and did not need over-wintering. Happily, our income did not suffer apart from the loss of stock, time and heating costs.

Geraniums were a great success. I had noticed in the catalogue reference to a new geranium named Sprinter, a new introduction from America described as a compact, bushy plant easily grown from seed. Sown in December and in flower by May it was similar to Paul Crampel but only available in one colour, others evolved later. It sounded too good to be true. It was expensive but I ordered some seed. Using the new propagator I sowed some in January which germinated quickly and came up like mustard and cress. We hurried to pot the seedlings into three inch pots which they stayed in until sold. They grew so rapidly we were constantly respacing them. Sprinter was well named - I've never seen seedlings grow so fast. In May they were in flower and selling as fast as we displayed them.

The next year we increased the numbers but were frantic to find space as they grew so fast and we even resorted to standing them between the tomato plants in the rows until it was safe to stand them in the open. Fortunately neither seemed to suffer. We improvised all kinds of temporary shelter. Empty mushroom baskets could be had for the asking and removing the pots we sold eight in a basket to the public, shops and garden centres. They were the best sale we ever had. At first we had a monopoly at Yarmouth but others soon cottoned on but never equalled our sales. We never fully overcame the space problem. There were too many plants!

Coleus were very popular and we only grew the finest selection, all brightly coloured, and by offering the best we built up a good trade. Harry Brunning, formerly well known for the shop in Regent Road and the nursery in Browston, urged me to try Cupheas Platycentra, the Mexican Cigar Plant. It had been out of fashion for years but he was sure it would prove a winner again. I was doubtful but tried it and he was right. It is a nice compact plant covered with attractive red flowers, similar to a fuchsia. I even tried some as border plants with great success.

Coleus and other annuals.

We aimed for a wide range of plants from spring to autumn. Wallflowers were a good line with mixed colours, sweet pea plants in pots were always in demand, also tomatoes and many other vegetable plants for gardeners. We tried seedling Christmas trees, eucalyptus and others. A sure seller was golden privet which, unlike the green variety, is not easy to strike. The secret is to strike the cuttings in a frame in July and you will have good plants the next year.

Being only in a small way of business we dare not plunge too deeply as horticulture is hard work with no sure return. Weather, fickle public taste and market forces are against the small grower. Our work was labour intensive but we could never afford the extra hands needed and casual help from neighbours, although welcome, was unskilled and could not be relied on week by week. We tried to plan our crops so we could deal with them in succession, mainly using our own labour. In certain seasons things ran into one another and it was a nightmare.

Small growers manage to survive by hard work, commonsense and thrift. You quickly learn never to throw anything away as it may come in handy and I accumulated a lot of junk in my time but it was amazing how useful it proved from time to time. We used a lot of string to support tomatoes, for instance, which we saved when they finished for bunching flowers. The cost of seeds can be reduced by saving your own. We saved runner beans, broad beans, wallflowers, sweet peas, sweet Williams and a few others. I would select the best, mark them, ripen and dry them using only disease free plans. At one time thousands of feet of ex-army telephone wire was sold cheaply on drums, very useful for supporting crops. Ex-army boots were new and cheap, also clasp knives which we were always losing. Government surplus tools came in very handy and sold well when we finished.

We made much use of mechanical aids, rotavators etc., but the finest tool ever invented was the 'Planet' hoe, an American invention, simple and effective, perfect for destroying weeds and keeping the soil friable and open. Pushed by hand, it was attached by two wooden handles to a metal frame holding two hoes or tines with one or two wheels, often copied but never bettered. The 'Planet' seed drill was another fine implement.

Posing with my new rotovator in September 1958.

When we started market gardening I only had hand tools and if a neighbour was willing and could spare the time he might horse plough the open land but they were not always happy to oblige as the branches and tree roots were a danger and a nuisance, even breaking a plough share. Later a friend, Charles Nichols, helped with his grey Ferguson tractor but it was obvious that more than a spade and fork was needed so I looked for mechanical help.

There was much conflicting advice. My interest was aroused by the Anzani Iron Horse, a garden plough used like a horse plough with an engine. Three speeds and reverse with a choice of ploughs and a cultivator included, all for £120 delivered, with other implements which could be added later. Scraping the barrel I found the money and ordered one from Harston's at Norwich. It duly arrived and was an excellent tool. The place was such a mess with masses of rubbish and the Iron Horse was easy to work and did a good job for an amateur like me. It was necessary to remove several trees and bushes to use it to the best advantage which I was loath to do as at that time we still relied mostly on fruit. Later when the fruit sales collapsed I was ruthless in removing anything in the way. It was a wonderful little machine and I improved its performance by buying a one-way plough when they became available which fully repaid the £26 it cost. The ploughing was so much easier and neater.

The workers, with our first tractor.

When we increased the acreage I bought a second-hand Ferguson with a cultivator from a man at Strumpshaw. I took Graham Cook with me and he drove my van home and I drove the tractor which had no licence or insurance but fortunately I met no policemen. I just about wore it out and then bought another grey Ferguson from Jack Newman of Fleggburgh. Billy Page sold me a double furrow plough cheaply because it had a broken share which my friend Percy Trett welded perfectly and I used it until I retired.

Working with tractors can be dangerous. While pulling tree stumps out with the Ferguson you could turn over in an instant if the rope was at the wrong angle. Removing trees was always done with the permission of my landlords, who were always sympathetic and helpful. The rented land was £10 for one piece and £20 for the other. Not much by today's values.

On one occasion the perils of the job were brought home to me

*Here I am shown ploughing in the second
of our tractors in November 1972.*

whilst ploughing with the Iron Horse. I was reversing when a branch caught the throttle and the machine leapt back pinning me against a tree trunk. With the wheels pressing hard I could not move but eventually I freed my arm and reaching the throttle managed to put the gear into neutral. It was worrying at the time but I survived and took good care never to get caught again. Unlike horses, tractors don't stop when you shout, "Whoa!"

❋　❋　❋　❋　❋

CHAPTER 6

Sitting the Market

When I first used to sit the market in 1947 I soon became known as the 'Apple Man'. A lot of work went into this and sometimes the money was poor. This was best summed up by Cecil Tooke, who caused a wry laugh one day when someone asked him who he worked for. "For three months I work for Barclays Bank and my overdraft," he said, "three months for Bessey and Palmer (Coal Merchants), three months for my father and brother and the rest of the time for myself and wife!" And that applied to most of us!

The expression 'sitting the market' is said to go back to the time when the custom was to throw your cloak down on the market place and spread goods over this covered area. Yarmouth market has a long tradition, and we were part of this tradition of small farmers and market gardeners selling home grown produce on the local market. However, the country stall holders are not

1854 print showing early stall-holders 'sitting the market'. In those days the market seats were made of basketwork and bore the occupant's name.

to be confused with market traders who go from town to town, or the Cheap Jacks and Flash Harrys whose slick patter and amusing antics enliven every market. All of course are stall holders but there was a subtle difference and it was accepted in my day. This was the time when Yarmouth market was mainly made up of country stall holders and the present generation has no idea of the wonderful variety offered. It was a hard and precarious living though, and unacceptable to young people today.

When I first knew the market there were hundreds of country stalls all offering their own home-produced fruit, flowers, vegetables, eggs, butter, pork, poultry, cream cheeses, brawns and wild and tame rabbits every Wednesday and Saturday, filling the market place from one end to the other in three ranks. At the south end were the six day stalls with fruit, chips, jellied eels, tripe and shell fish and the well-used tea stalls.

Chip stalls first appeared around 1902 and Yarmouth market from then on became famous for its chips with well known fryers such as Brewer, Thompson, Kelly and Nichols still around today. How many people, I wonder, can remember the old Yarmouth rhyme?

"North, South, East or West,
Yarmouth girls they are the best,
'cause when you kiss their greasy lips
they always taste of Yarmouth chips!"

Other types of stall were fitted in if there was room but the great majority were country producers from the surrounding districts. On Saturday night every stall, including the six-day stalls which were mostly on wheels, had to be removed and the Market Place left as an open public space. Nothing permanent had been allowed since the days of the pillory and the Market Cross. The present expensive intrusions are a typical municipal extravaganza. The market, throughout its long history, was profitable to King or council. Now due to the muddled policies it seems more of an incubus but it is still the heart and hub of Yarmouth. Sadly the country stall holders

The Thompson brothers with their first chip barrow (above)
and their first stall (below).

Yarmouth market circa 1900.

who contributed so much to the attraction of the market have dwindled to a mere handful. The country folk mostly came in horse-drawn market carts or trollies which they parked on Hogg Hill, Theatre Plain and other spaces, the horses being stabled in pub yards, contractors' stables and some in the Rows. They set out early in the mornings and returned late at night and it is said they frequently fell asleep but the horses knowing the way brought them safely home! A story is told how one stall holder unable to finish plucking her poultry at home continued plucking all the way to Yarmouth leaving a trail of feathers behind!

Until about 1930 the market was let to the highest bidder who then recovered his rent and profit from stall rents and charges for pitches on the Fair. Mr Cosh Beavor was the last to do so and Mr Tooley was his superintendent. After that the Corporation took it over and appointed Mr Don Davey, Tooley's nephew in his place. He was outstanding and ran the market in the best interests of stall holders and Corporation alike. He was always fair but firm and well liked and respected by all. He joined the R.A.F. but resumed his duties after the war. During his time the market was well run and profitable although rather untidy looking which always annoyed the Council which yearns for nice tidy looking, shiny plastic sorts of stalls and plastic people without any individuality. Don was well known through his speedboat racing on Oulton Broad and other places.

For years there had been arguments about the legal position of stall holders *vis-à-vis* the Corporation. To clarify the situation on 1st March 1947 the Corporation served notice terminating stall holders' occupation of their stall, site or pitch and requiring them to re-apply under conditions set out by the Corporation. Stall holders had been said to have prescriptive rights but it had all been very vague.

At one time the people of Ormesby claimed freedom from tolls at Yarmouth and things came to a head in 1862 when a court case, 'Groom versus the Corporation', finally decided the matter and ruled that whilst the people of Ormesby did not have to pay toll they still had to pay rent as the Corporation owned the 'soil' of the market place. This supported Yarmouth in a later dispute with the County Council over roads in the town.

The Easter Fair on the market place in 1900.

An important event on the market place was the annual fair which took place following the great Easter Fair at Norwich. This arrived in Yarmouth the following Wednesday and began to set out its stands as the market stalls were removed. It was fascinating to watch the speedy erection of the large and complicated amusements to such high standards. I cannot remember any accidents except from foolish behaviour. Before the war the crowds were so large that many shops around the fair erected boards over their windows to prevent the press of people pushing them in. The fair was an important day in the year, a real treat. Some folk spent more than they could afford and a certain amount of good-natured horseplay took place, much confetti was put

Yarmouth's last Market Cross was removed in 1836 and not replaced.
These structures were common features of market places at one time,
symbolising the community's strong Christian brotherhood.
A cruciform stone now marks the site instead

Rare photograph of the market pump which was situated opposite the
'Two Necked Swan' public house. Water pumps were vital at one time
as they were the only source of clean water for many people.

The market was also the setting for public meetings, parades and demonstrations. Here crowds gather for Queen Victoria's diamond jubilee in 1897.

A photograph, dated 1898, of a Wednesday market - a Saturday market would have had more stalls. Every building on the southern and eastern side of the market place, that is every building pictured here, has been demolished. The Fish-Stall House was the last to go in January 1974.

A busy market in the mid 1860s. On the left of the photograph can be seen ▶ an example of the Yarmouth troll cart which were specially designed to go through the Rows.

Yarmouth market in 1864. On the right can be seen the entrance to the fish market which was erected in 1844. The selling of fish on Yarmouth Market is still forbidden, a tradition which dates back to this time, as the two markets didn't want to be in competition with each other. Of course, the fish market has long since closed but the tradition lives on.

down girls' dresses and so on. Local bakers used to make 'fair buttons' and Watson's still do. After eleven on the Saturday night everything had to be dismantled and removed and away the showmen went for another year. Such colourful characters; the Stocks, Grays, Manners, Thurstons, Bells and all the rest, great individualists yet united in the Showman's Guild which upheld standards and jealously guarded their rights and interests.

In the days before radio and television the market place was the setting for public meetings, parades and demonstrations. Crowds gathered to hear and heckle the politicians and other speakers and the market place has witnessed some remarkable scenes. Anthony Fell M.P. was the last I heard there, a good speaker. At one time the local M.P. would visit the market, stopping to crack a joke here and there and getting the feel of things. Strangely, prospective members of parliament and councillors usually appeared just before elections!

Market trade was much diminished during the war years as all school children and many families were evacuated to safer areas. Things improved in the latter part of the war but local growers sent the bulk of their produce to London and the bigger markets. Production was strictly limited to essential foodstuffs, and transport and distribution closely controlled. A good many stories could be told of how these regulations were craftily evaded and certain 'luxuries' found their way to the bigger markets and higher prices.

I have to emphasise the long hours and hard work; this was the lot of all market growers and accepted as part of our lives. In the busy times I got up at 4 am, made tea and took a cup up to my wife and then drove our van, loaded the previous night, to market. On arrival I'd fix the canvas covers and stack the boxes of produce under the stall and then return for a second load which would then be stacked up at the back of the stall. By this time it would be near 6 am.

Returning for the third load I would check the chickens and green-houses then have breakfast, load the van and we both returned to the stall where we quickly set out our produce with side tables for flowers or plants etc. all ready for customers. With scales, weights and cash box all at the ready I left Kathleen in charge and returned for the fourth and final load. At home the post was opened in case of orders or anything urgent, a quick check of the greenhouse ventilation and I then finally loaded and returned to market for the day.

At mid-day my wife would complete her shopping and return home by bus. The main rush was from 8 am to 11.30 am after which I could cope on my own. On arriving home my wife would immediately check the greenhouse ventilators and the poultry as so much could happen in a few hours. Poultry and growing plants are very vulnerable and need close attention. In hot sunshine temperatures rise rapidly and cold winds play havoc if unnoticed. At the stall the morning trade was vital. If the bulk of our produce was not sold by mid-day we were in for a long day. In the busy time we hardly had time to speak to each other and it was good to see our produce disposed of so rapidly. Most of our regular customers knew what they wanted and wasted no time. Passing trade had to take its turn and people probably thought us brusque at times but the regulars had to come first, they were our bread and butter. After mid-day things were easier. I seldom stayed

At our stall during the cactus craze. My old van can be seen in the background in the days when stall holders were allowed to keep their vehicles at the rear of their stalls.

after 5.40 pm as there was so much to do at home, from glasshouse watering to feeding the chickens.

For many years we were allowed to park our vehicles behind our stalls which was very convenient. On arrival at market we first fitted the cover over the stall and anchored it with heavy weights, especially in windy weather, next we set out the produce in trays and flowers in buckets all ready for customers. Prices were the next concern. What should we charge? The bigger growers set the price and we usually fell into line. Supply and demand governed the wholesale markets and prices fluctuated greatly. A lot depended on quality. It was hard to decide the right price as a penny or two either way made a big difference at the end of the day.

We would stroll up and down the ranks quizzing prices. Some were crafty and displayed high prices at first only to reduce them as customers arrived. We would notice people passing us by and realising something was amiss adjust prices accordingly. I always believed if you had a good crop, reasonable prices and quick returns were the best policy in local markets. It was very difficult with over production and low prices as most of the produce was perishable. Gluts were frequent and much was thrown away. Billy Page used to say 'scarcity is the growers' only friend' which may not be entirely true but it was strange how buyers increased their orders when things were scarce! When prices had a sharp drop growers described them as 'Humpty Dumpty'.

The market was a busy, bustling place, full of character. Bob Savory was well known and was said to have started by selling fruit from a barrow on the seafront and went on to build a wholesale and retail fruit and vegetable business supplying a wide area. The warehouse was beside the market with its banana ripening chamber underneath and the family living accommodation above. Old Bob was fat, ponderous and rather untidy but young Bobbie was slim, brisk and neat with his hair plastered down in a parting. I never saw him wear a hat. They also had a big stall on the main market crossing and there Mrs Savory and her daughters, Marjory and Vera, reigned among the fruit and vegetables.

The Bartram's have had a long association with the market and here we
see how their stall has looked through the years. The top picture shows
the family stall in 1946, when it was on wheels, with Mrs Bartram on the
right, while the bottom view shows the stall in 1993 in the new undercover market.
Here Mrs Bartram is flanked by her sons, Ronny and Tony.

McCarthy's stall. On the left is Mrs McCarthy who sadly died in 1998 aged 94. Dan McCarthy is on the right.

Paul Richards at his father's stall in the early 1970s. Dave Richards had two cloth and material shops in King Street in addition to his stall on the market. Sadly both father and son have since passed away.

Almost adjoining the Co-op and near Savory's was Dan McCarthy, another fruit and vegetable wholesaler. Dan McCarthy was a flamboyant character, said to have come to Yarmouth from Penzance on a herring drifter looking for work. Like Bob Savory he started in a small way and by sheer hard work and ability built up a big business which is even bigger today. George Brewer was manager for several years. The names of Bartram, Roland, Weldon and Eagle with their six-day fruit stalls spring to mind, Tommy O'Connor's tea stall with his guitar and his book about the market and his other experiences, and there was old George Mann and his tea van. He was a cantankerous old so and so and when the council proposed changes he would not conform and retired.

Well known were Mr and Mrs Youngs from Rollesby and their son Freddie. They were typical small holders but were distinguished by wearing buskins, those highly polished leather gaiters then favoured by many countrymen. Farmer 'Shacky' Bond of Belton always wore a bowler hat and buskins when he visited the markets, things we seldom see today.

Mr Freddie Hubbard of Burlingham, nicknamed the 'Raspberry King' by Dan McCarthy, remembers Acle New Road before it was tarmacked, all terribly rutted and pot-holed, bad enough to upset the carts and in frosty weather it was necessary to put studs in the horses' hooves to prevent them slipping.

He also recalls the old custom 'meat for manners'. This was when at the end of the fishing season local haulage contractors arranged with local farmers and market gardeners to take some of their horses for the winter and use the animals and care for them at a time they needed an extra horse but the contractors had no work for them. They were fed (meat) and kept in working condition at no charge which suited both parties.

A colourful character was Freddie Broom with his straw hat festooned with big bunches of small tomatoes. Mrs Ward with her black and white cherries and Robin pears was from Filby. She lived in the 'Cherry Orchards,' a property dating back to 1600 with cherry trees older than memory. Little Charlie Boulton did a brisk trade at

Christmas with his hundreds of holly wreaths. Teddie Thurlow with his wife and daughter had a big country stall and sold enormous numbers of Christmas wreaths. Few people today realise the long hours, hard work and expertise that went into making them in the short time while the holly was still fresh.

There was Mr Nichols and his daughter from Acle offering skinned rabbits, wild or tame, and their lovely brawns. My friend Ernie Morse from Belton was a man of great integrity with an encyclopaedic memory of local people, events and old gardening practices of years gone by. His father was one of those specialist celery growers who cultivated the low-lying boggy lands of the Haddiscoe, Thurlton and Belton area. It was mostly hand work carried out by specialist teams. It was grown on baulks, well moulded up and they grew first-rate celery. These old ways and low lands have long since gone out of use and I doubt if anyone living can recall the details today. Most celery now comes from the Fens.

For years the Beares of Browston and Belton, a well-known gardening family attended Yarmouth market. Sam Beare, the finest grower of them all, was a grower wholesaler whose speciality was arum lilies at Easter. Another great character was Jack Howlett from Fleggburgh who started with his mother and built up a big business after the war buying and selling wholesale with his son Tony. They had a popular flower stall which is still on the market. He had a great sense of dry humour with off-the-cuff remarks and humorous asides, and his quiet generosity will long be remembered.

Lexie Dove was well known for her water-cress, bunches of snowdrops, daffodils, blackberries and many country odds and ends. Some well-to-do ladies thought she was hard up and gave her all sorts to help her out, little thinking she would leave a small fortune.

'Pal' Tennant of Scratby was renowned for his strawberries. Sadly he was killed by a traction engine, one of Thurtle's, at Ormesby while pulling out a hedge. Cranes of Ormesby specialised in organic vegetables, especially excellent tomatoes.

Everyone knew the Tooke family, old Archie and Mrs Tooke with their sons Cecil and young Archie. They were particularly known for high quality tomatoes and their outstanding chrysanthemum blooms.

A scruffy little man using hardly more than a board beside someone's stall used to shout, "Fowls are cheap," and he was usually referred to by that name. He went to Acle sale and begged all the birds condemned by the auctioneers as unfit for sale, dressed them and offered them at 1/- or 1/6d. Such practices are not allowed today, thank goodness.

Mrs Goose came from Martham. She lived to be a hundred and her daughter Violet Cossey carried on the stall. Tom Curtis was a real jolly countryman from Fleggburgh. Mrs Long of Scratby was well known with her Dr. Harvey apples. She lived in possibly the oldest cottage, a long low clay lump building with a very old orchard.

Mrs Sharman Wright also came from Scratby and she and her daughter Dolly Haylett were well known. Mr Sharman Wright was reputed to always have the earliest English new potatoes in the market years ago. From Ormesby St. Michael came Miss Nichols and her brother Andrew, also the Notley brothers with their flowers, grapes, fruit and vegetables. Another great character was Mrs Browne of Martham. Mr and Mrs Baldry came from Burgh Castle with a fine selection of produce from their nursery.

Humphrey Walpole from Filby was well known for his apples. He had a very old orchard but as trees died he kept replacing them with new varieties so he had a wonderful mixture. He was secretary of the local N.F.U. horticultural branch and also a music teacher. He lived in an interesting, very old, half-timbered clay lump cottage and had rights on Filby Broad.

What a profusion of fresh country produce all these stalls offered but we must not forget the others. Very popular were the pork stalls with their succulent portions of the pig, especially sausages. There was Chapman of Martham, Charlie Fuller of Filby, Jimmy Smith of Ormesby, Prime from South Walsham and Lennie Allen of Hemsby. Allen's always had a queue. So famous were their sausages, if you did not order you did not get any. They were really tasty, not like the poor substitutes of today!

These are only a few, many I forget, but I well remember some other market characters. During race week a well-known racing tipster, Prince Monolulu, whose catch-phrase, 'I've got a horse,' was heard at many a race-meeting, was a great attraction. A coloured man dressed as an African chief, he visited the tea stalls and even bought William pears from me as at one time it was customary for race-goers to buy pears for the September meeting.

Another attraction was Alf the 'Purse King' whose patter always drew the crowds, especially when Gus Elton the famous comedian from the Wellington Pier called at his stall which he frequently did. The exchanges between them were a show on their own. Alf had a big turnip watch on a large chain secured to his waistcoat by a big padlock and Gus pretended to steal it. How they drew the crowds, which did Alf's stall a power of good. Another crowd puller was David the crockery man with his slick patter while juggling with his wares. A very popular stall was David's.

The Calvers from Thurlton had a poultry stall. Old Mr Calver had lost an arm in the First World War and in its place he had a hook, something we do not see today, and it was marvellous what he could do with it. His son George and his wife carried on the business. Dick Grint came from the same area with eggs, celery, game in season and other country produce. I almost forgot dear old 'Soldier' Sims from Filby who had served in the Great War.

Among the drapery and clothing stalls were two men from Leicester with a great selection of factory seconds in pullovers and jumpers, and another couple were two very smart oriental traders with splendid turbans. One popular stall was Heywood's stocking stall. They had relations with a factory in Leicester and sold their seconds or slightly imperfect goods.

Billy Simmonds from Runham Vauxhall had the finest selection of bedding plants anywhere. He was most particular and any plants not up to standard he threw out. His wife and daughter helped him on the stall until acute arthritis forced him to retire.

Of all the people we saw in the market there were two we referred to as the 'Big Drips' and our hearts sank on seeing either of them approach as they both had drops at the end of their noses of which they were completely unaware. They came separately and were very old. When either stopped at the stall the drop grew larger as the drip descended. What could be done without hurting their feelings?

Two well-to-do ladies regularly patronised our stall and we were delighted to serve them until another stall holder said, "You need to watch them, they take more than they pay for." We were shocked but it was true.

Another time I was busy at the back of the stall and out of the corner of my eye I saw a man take some of our William pears and put them in his bag. In a flash I spun round. "Put them back," I shouted. Without a word he meekly put them back and offered me a shilling which I promptly took although he had no pears! One winter there was heavy snow about Christmas. The council had cleared the snow from the stall sites but left it piled around the market and our van was lodged in this snowdrift. It was useful as we stuck our Christmas trees in it for display. At the end of the day I had two left and as I was loading the van they vanished, some smart thief took them. Even in those days we had to watch out when busy.

We had annual visits from the Little Sisters of the Assumption collecting for their devoted work in Norwich, their happy, saintly faces inviting support. Not so another, dressed like a nun, who looked far from saintly and whose credentials I questioned. She produced a dubious, scruffy handwritten document and I reported her to the police but she disappeared. Maybe I was too suspicious - who knows?

I particularly remember one flamboyant character who often visited the market. He had a peaked naval cap, a silk scarf, a sash round his waist, his trousers were stuffed into leather boots and he looked like an Italian organ grinder. He travelled miles round Gorleston, Yarmouth and the surrounding villages. For years he pushed a companion whom he referred to as 'Mother' around in a wheelchair but whether she was his mother or his wife I never did discover. He sometimes hung a tray round his neck displaying all manner of odds and ends. At one time he had a detailed model of a village and using a cigarette he made smoke come out of the chimneys. He lingered one day in the market so I asked if I might take his picture as I had our little Brownie camera at the stall. He reluctantly agreed provided he received a copy of the snap. A few weeks later he appeared looking really ill and asked for the photograph, saying he had been in hospital. Unfortunately I had not finished the film. and sadly we never saw him again although we kept the snaps for him.

Many of the stall holders were characters in their own right too. Mr Spinks, a small grower from Belton, was situated next to our stall on the market. He was an ex-London policeman or fireman, a friendly and helpful man, latterly succeeded by Aubrey Roach who had a great sense of humour. Aubrey would note passers-by and assign them a role in life. As a certain lady passed he would say,

"There is Miss Cunningham-Brown just back from the Amazon and the wild Indians!" The description would fit exactly! When trade was slow and the bargain hunters were prowling the market Aubrey would call out, "Come on, you vultures!" He grew a lot of dahlias in competition with my chrysanthemums and as we compared them, unthinkingly I said, "Any fool can grow dahlias," but he took it in good part. Shuckford the pork butcher had a quick sense of humour too. One day as he was putting a pig's head into a woman's basket she said, "I see you've left its eyes in." Like a flash he replied, "Yes, so it can see you through the week!"

Here I am again, at our stall with Ernie Morse (left) and Mr Stevens (right) in the background.

77

In 1927 Mr Harvey (pictured right) started his bacon stall on the market. Mr Harvey tells me that he made this stall himself and it was the first to have wheels. In the early days he would use the telephone in Foulshams eating house and they indicated that there was a call for him by blowing a whistle. Of course, many of his customers used to think he was going for a drink!

By the 1960s Harvey's bacon stall looked rather different. It closed in the mid-1980s and is still missed by many.

The Brewer family at their hot peas and jellied eel stall, dressed for the Coronation of George VI. At this time they had a chip stall and a fruit and veg stall adjacent to one another.

The view from the other side of the stall. Dick Underdown is shown on the left busy at work with his customers.

My stall, situated on the west side of the market, was near Woolworth's and you can see me over the shoulder of the woman browsing at the front.

A Wednesday market in May 1971. The market was divided in two by a crossing which can be seen on the photo. The stalls to the south of it were there on a semi-permanent basis and remained there all week except Sundays. Late on Saturday night they were dismantled, or in the case of chip stalls that were on wheels, pulled away, as the market had to be clear on Sunday. The stalls to the north of the crossing were country produce stall and those of the 'cheap jacks'.

The view from our stall, looking towards St. Nicholas Church.

CHAPTER 7

Doin' Different

Compared with other occupations ours was an uncertain and hard life but I do believe hard work and sacrifice, with a purpose, is worthwhile if you enjoy what you are doing. We were always busy, interested, happy and contented. However difficult, I always found time to take part in village life and wider affairs as an awkward independent, always prepared to stand up and be counted, pursue the truth and keep at it, however unpopular in certain quarters.

After we got over the early troubles, although we were hard-up compared with regular wage earners, I endeavoured to play a part in local council and church affairs and also more widely. I was thrilled by the coming to East Anglia of the new university and we contributed to the appeal for funds and continued as supporters for the next twenty years. We grew to love Norwich Cathedral and became Friends from about 1960 onwards. In 1968 we were founding members of the Parson Woodforde Society and attended their frolics when possible. Another cause to attract our support was the restoration of the Fishermen's Hospital in Yarmouth and we contributed to that.

We could not really afford any of the these efforts but felt ordinary people should do their bit. We had no television, only radio. When the operas came to Norwich I managed to persuade Malcolm Freegard to give me a few minutes on BBC Television and that paid for the seats! In one way or another we did our best for Norfolk and felt it well worth the sacrifices.

BBC Radio was a happy companion when potting and doing other jobs in the greenhouse or during the many hours spent tying flowers.

The programmes then were so full of interest, a varied choice of tuneful music and so many marvellous entertainers we would not miss for worlds. We have never had a television set as we have never had the time to waste.

Like most people we had great affection and respect for the BBC in those days and I felt privileged to take part in several programmes on radio and television. It began in 1962 when I was trying to interest people in preserving the old apples which still survived, in particular the well loved Dr. Harvey, so popular in East Anglia for over 350 years.

I wrote to BBC Norwich suggesting that they took up the matter, thinking they had crowds of eager young reporters straining to rush out and investigate. To my surprise a letter arrived saying how interested they were and inviting me to the studio to discuss the matter further. It was not easy to spare the time but I duly arrived at All Saint's Green to be welcomed by attractive June Mansfield sedately floating down that splendid staircase to conduct me to David Bryson's office upstairs. He was responsible for establishing the BBC in Norwich. He asked me all about the old apples and then scared me to death by inviting me to talk about them on radio and television. I was petrified but felt I had to go through with it for the sake of the old apples. Paul Tilley was the producer on radio and I think the T.V. producer was John Bacon. Malcolm Freegard was head of BBC television and was both interested and enthusiastic, soon making me feel at home. It was terribly hot under the lights in the studio in those days and the make-up people made us look presentable. As far as I know the programme went over fairly well, the producer saw to that.

My next appearance was with Ted Ellis. I did three programmes with Ted, recorded in the afternoons and re-broadcast at 8 am. I greatly enjoyed these programmes but had give up this interesting introduction to broadcasting as I could not afford the time. It was a shared programme and we only received £3.3.0d which was £1.11.6d each. This was acceptable to Ted who only had to walk from the Castle Museum in his lunchtime while I had to give up half a day's work and travel to Norwich from Ormesby, so sadly I gave it up.

However, over the next ten years I made several appearances on 'Look East' and broadcast on radio from Bristol, Birmingham and London. I felt proud to broadcast live from the famous Broadcasting House in London. I was three times live on 'Woman's Hour' and three times live on 'Home this Afternoon'. I felt honoured to give a thirty minute broadcast on Parson Woodforde in a series on great diarists and have tea with the Deputy Director of the BBC later. I had requested a visit to the Director General but it was unlikely he would receive such a small fry and I was lucky to meet his deputy! Like many listeners I was not happy about the direction programmes were taking and wrote straight to the top. I was invited to the inner sanctum to meet the big Chief's deputy whose name I cannot recall. He was most charming and his secretary trundled in a tea trolley with nice cups and saucers, a tea pot etc. and some first class biscuits, not the plastic cups and dry old biscuits offered to lesser beings!

He was retiring soon to a really delightful place in Suffolk called Bildeston which I knew well and he artfully got me talking about fruit he hoped to grow and then said how delighted he'd been to meet me and what was it I wished to discuss? Oh, the crafty so and so, so pleasantly passing the time on fruit. Taking the bull by the horns I told him many listeners were dismayed by the lowering standards in programmes and we felt they were pandering to the lower tastes, also we objected to bad manners and bad language all too frequently seen and heard on many programmes. He explained that the BBC had given much thought to this and considered it was their duty to present the world as it was seen and heard by most people and not as we might like it to be, otherwise it would be artificial and unreal, much as we might wish for an ideal and nicer setting. I replied we felt the emphasis was wrong as we had to receive this stuff in our homes where we did not allow such language or behaviour. He said he regretted it as much as I did but we had to live in the real world. He thanked me for my comments and there we had to leave it as my time was up. Altogether I had a very interesting day visiting several studios and departments, and the BBC paid my expenses and a fee.

I have never been very impressed with showbiz and media personalities and some of their antics seem childish to me. On one occasion I was showing an injured seagull on television to draw attention to the suffering caused by irresponsible cowboy marksmen on Breydon Marshes and a well-known theatre personality on the programme said when he spotted the gull that he was not competing with animals. The producer calmed him down but I could see his point - the gull was far more interesting!

I did not like scripted recordings. It was like being in a glass case, the producers were outside and it was impossible to judge how it was going, they looked so serious and off putting. Twenty years later I was much happier with Radio Norfolk and for twelve years Peter Glanville, Keith Skipper and other producers summed me up and allowed me loose on the air. I am told my voice is distinctive and people would ask where they had met me only to discover it was on the radio. Some said they would know that voice anywhere which I found flattering.

In recent years I have turned to writing books. Here I am signing copies of 'I Remember Yarmouth' at Jarrolds in 1995.

'Look East' showed our stall, then the gourds and other oddities at home. The best was John Seinfield from Anglia TV showing the last Dr. Harvey's on our stall and then me felling the last Dr. Harvey tree in 1970 as the end of an era. To end the show he took a bite out of one of the apples. "Delicious," he said but how he managed to smile I don't know, the apple was far from ripe! I still have the sound recording which was presented to me. It was all very interesting but the fee was so small it was hardly worth the time and trouble.

In 1979, with my 65th birthday imminent, we decided to retire and enjoy our other interests. Scratby Garden Centre heard we were selling all our greenhouses and other equipment, came to see it and made a very satisfactory offer for the lot which I accepted. As part of the deal I offered to dismantle, repair, paint and re-erect the greenhouses on their prepared site. It proved to be a bigger undertaking than I expected and I was offered payment as the work progressed but a deal is a deal and I kept to my promise. The houses did well for a time but the bottom fell out of horticulture and unfortunately they did not enjoy the full benefit of their venture.

We finished going to market but continued growing and selling chrysanthemums to Yarmouth florists for about three years then we finally retired. Some people tend to look down upon those who go to market and I was amused when my better-off relations were not happy to be seen at our stall, calling me ' The Peasant'.

We were happy though and in my life I was a giver, not a getter and a doer, not a viewer and the people of Ormesby must have respected my integrity and care for their interests because to our surprise, they made presentations and thanked us at a public meeting when we left Ormesby. We were not afraid of hard work, enjoyed good health and a love of independence. We sometimes struggled to survive but achieved a modest living in the end by choosing to 'Do Different' as we say in Norfolk.

✳   ✳   ✳   ✳   ✳

Jim Holmes, Apple Man